Fifteen Animals on the Farm
15 farmyard animals, and counting!

P.J BARNES

Copyright © 2016 Philip James Barnes

All rights reserved.

ISBN: **1534643761**
ISBN-13: **978-1534643765**

CONTENTS

No Animals on the Farm   Pg 2
1 Animal on the Farm   Pg 4
2 Animals on the Farm   Pg 6
3 Animals on the Farm   Pg 8
4 Animals on the Farm   Pg 10
5 Animals on the Farm   Pg 12
6 Animals on the Farm   Pg 14
7 Animals on the Farm   Pg 16
8 Animals on the Farm   Pg 18
9 Animals on the Farm   Pg 20
10 Animals on the Farm   Pg 22
11 Animals on the Farm   Pg 24
12 Animals on the Farm   Pg 26
13 Animals on the Farm   Pg 28
14 Animals on the Farm   Pg 30
15 Animals on the Farm   Pg 32

NO ANIMALS ON THE FARM

There are no animals on the farm,
The farmer just can't find any,
He needs help so he sounds the alarm,
Hoping you can find and count many!

How many animals are on the farm?

0

There are NO animals on the farm!

Shall we keep counting for more?

1 ANIMAL ON THE FARM

Let's have a look and see what we find.
Oh look, is that a little Mouse peeping?
But what about her farmyard friends,
Do you think they're all still sleeping?

How many animals are on the farm now?

1

There is ONE animal on the farm!

Shall we keep counting for more?

2 ANIMALS ON THE FARM

The Farmer heard a 'Baaa' from a little Lamb,
Then spotted her white fluffy wool.
How will the Farmer get out of this jam,
And return his farm to being full!

How many animals are on the farm now?

2

There are TWO animals on the farm!

Shall we keep counting for more?

3 ANIMALS ON THE FARM

From distance there is a loud 'Cluck Cluck',
The Chicken has come back home,
It seems the Farmer has found good luck,
But where do all the other animals roam?

How many animals are on the farm now?

3

There are THREE animals on the farm!

Shall we keep counting for more?

Moo, says the big friendly Bull,
Happy to be back with his friends,
However the farm is still far from full,
So this is not where the story ends!

How many animals are on the farm now?

4

There are FOUR animals on the farm!

Shall we keep counting for more?

5 ANIMALS ON THE FARM

Waddling in, the Duck says Quack, Quack!
Happy and splashing in a puddle,
It seems the animals are all coming back,
But with only five, we're still in a muddle.

How many animals are on the farm now?

5

There are FIVE animals on the farm!

Shall we keep counting for more?

6 ANIMALS ON THE FARM

Ah, in flies the white farmyard Goose,
Back from a day of exploring,
Yet there are still more animals on the loose,
We better find them before rain starts pouring!

How many animals are on the farm now?

6

There are SIX animals on the farm!

Shall we keep counting for more?

7 ANIMALS ON THE FARM

What's that, standing beneath the apple tree?
It's a Goat who lets out a loud bleat,
We should look for more, if you agree,
With seven in total the farm is close to complete!

How many animals are on the farm now?

7

There are SEVEN animals on the farm!

Shall we keep counting for more?

What do you see, over by the big barn doors?
It's a little doggy, barking 'Woof, Woof'
I think we should keep looking for more,
To get all the Farmer's animals under one roof!

How many animals are on the farm now?

8

There are EIGHT animals on the farm!

Shall we keep counting for more?

9 ANIMALS ON THE FARM

Hopping into the farm is a grey bunny rabbit,
Listening to all of her friends with big ears,
Counting more animals is becoming a habit,
We're helping to calm the Farmer's fears!

How many animals are on the farm now?

9

There are NINE animals on the farm!

Shall we keep counting for more?

10 ANIMALS ON THE FARM

Shush, what is that you hear?
I think it's a miaow from a Cat,
On every page we see another animal appear,
Now we have ten to look at!

How many animals are on the farm now?

10

There are TEN animals on the farm!

Shall we keep counting for more?

11 ANIMALS ON THE FARM

Oink, Oink! In comes a cute little pig,
Happy with a curl in her tail,
Our farm is starting to get really big,
To count all 15 Animals, we will not fail!

How many animals are on the farm now?

11

There are ELEVEN animals on the farm!

Shall we keep counting for more?

12 ANIMALS ON THE FARM

Looks like we have found another to add,
A big grey donkey screaming Eeyore!
The farmer is going to be really glad,
Let's help him find just three more!

How many animals are on the farm now?

12

There are TWELVE animals on the farm!

Shall we keep counting for more?

13 ANIMALS ON THE FARM

What kind of animal says Gobble, Gobble?
It's a big bird called a Turkey,
This one came walking in with a wobble,
Because he was filled with glee!

How many animals are on the farm now?

13

There are THIRTEEN animals on the farm!

Shall we keep counting for more?

14 ANIMALS ON THE FARM

With great presence, in trots a big brown horse,
Letting out an almighty neigh,
Almost all of the farm is out in force,
We just need one more to complete the display.

How many animals are on the farm now?

14

There are FOURTEEN animals on the farm!

Shall we keep counting for more?

The farmer yells out Woo-hoo!
He has just seen his Rooster come home,
Can you hear a loud 'Cock-a-Doodle-doo'
To celebrate not one animal being alone!

How many animals are on the farm now?

15

There are FIFTEEN animals on the farm!

Great job! We can now stop counting for more!

ONE ANIMAL ON THE FARM

You are a great counter!
Now that you can count all 15 animals,
Do you want to try counting the below?

How many Farmers are there?
What about Barn doors?
How many legs does the horse have?
How many horns does the Bull have?
How many Trees are there?
How many Apples can you count?

Shall we keep counting for more?

THE END – HOPE YOU HAD FUN!

www.ingramcontent.com/pod-product-compliance
Lightning Source LLC
Chambersburg PA
CBHW050357180526
45159CB00005B/2058